NUREG-0654
FEMA-REP-1, Rev. 1
Supplement 3

I0467997

Criteria for Preparation and Evaluation of Radiological Emergency Response Plans and Preparedness in Support of Nuclear Power Plants

Guidance for Protective Action Strategies

Manuscript Completed: October 2011
Date Published: November 2011

ABSTRACT

Supplement 3, "Guidance for Protective Action Strategies," to NUREG-0654/FEMA-REP-1, "Criteria for Preparation and Evaluation of Radiological Emergency Response Plans and Preparedness in Support of Nuclear Power Plants," updates the previous version of Supplement 3, "Criteria for Protective Action Recommendations for Severe Accidents," issued July 1996. Supplement 3 provides a protective action strategy development tool based on recent technical information and is intended for use by nuclear power reactor licensees to develop site-specific protective action recommendation procedures. Offsite response organizations should use Supplement 3 to develop protective action strategy guidance for decision makers. The guidance of Supplement 3 provides an acceptable method to comply with Appendix E to Part 50, Title 10 of the *Code of Federal Regulations* (10 CFR) Section IV, paragraph 3 in the use of evacuation time estimates in the formulation of protective action recommendations (PARs) for the plume exposure emergency planning zone, and provides guidance for the provisions of 10 CFR 50.47(b)(10) in the development of a range of PARs. Supplement 3 also provides guidance to support the information in NUREG-0654/FEMA-REP-1 that the U.S. Nuclear Regulatory Commission finds to be an acceptable method of meeting the requirements in 10 CFR 50.47(b)(7) for the development of a public information program. However, licensees may identify alternative methods of compliance with these requirements.

PAPERWORK REDUCTION ACT STATEMENT

This NUREG contains and references information collection requirements that are subject to the Paperwork Reduction Act of 1995 (44 U.S.C. 3501 et seq.). These information collection requirements were approved by the Office of Management and Budget, approval number 3150-0011.

PUBLIC PROTECTION NOTIFICATION

The NRC may not conduct or sponsor, and a person is not required to respond to, a request for information or an information collection requirement unless the requesting document displays a currently valid OMB control number.

CONTENTS

Page

ABSTRACT iii

1.   INTRODUCTION 1

1.1   Purpose 1
1.2   Regulatory Basis 1
1.3   Protective Action Guides 2
1.4   Background 3

2.   PROTECTIVE ACTION STRATEGY GUIDANCE 7

2.1   Development of Site-Specific Protective Action Recommendation and Decision Logic ... 7
2.2   Notification of Protective Action Recommendations at a General Emergency 8
2.3   Termination of Protective Actions 8
2.4   Precautionary Protective Actions at Site Area Emergency 8
2.5   Wind Persistence Issues 8
2.6   Expansion of Initial Protective Action Recommendations 9
2.7   Strategy for Rapidly Progressing Scenarios 9

3.   PUBLIC INFORMATION MATERIALS 11

3.1   Public Information 11
3.2   Emergency Messaging 12

4.   GLOSSARY 15

5.   REFERENCES 17

ATTACHMENT: Protective Action Strategy Development Tool A-1

1. INTRODUCTION

1.1 Purpose

This guidance is provided for use in developing site-specific protective action strategies for implementation during an incident that escalates to a General Emergency at a commercial nuclear power reactor site.[1]

1.2 Regulatory Basis

Appendix E to Part 50 of Title 10 of the *Code of Federal Regulations* (10 CFR) Section IV, paragraph 3 states, in part, "Nuclear power reactor licensees shall use NRC approved evacuation time estimates (ETEs) and updates to the ETEs in the formulation of protective action recommendations...." Section 50.47(b)(7) provides, "Information is made available to the public on a periodic basis on how they will be notified and what their initial actions should be in an emergency (e.g., listening to a local broadcast station and remaining indoors), the principal points of contact with the news media for dissemination of information during an emergency (including the physical location or locations) are established in advance, and procedures for coordinated dissemination of information to the public are established." Section 50.47(b)(10) provides, in part, "Guidelines for the choice of protective actions during an emergency, consistent with Federal guidance, are developed and in place...."

This supplement contains guidance on new Part 50, Appendix E, Section IV, paragraph 3; additional guidance on the public information program required by 10 CFR 50.47(b)(7); and changed guidance on the formulation of PARs under 10 CFR 50.47(b)(10). Applicants and licensees[2] may voluntarily[3] use the guidance in this document to demonstrate compliance with Appendix E to Part 50, Section IV, paragraph 3 and the applicable provisions of section 50.47(b)(10) and, when used in conjunction with related parts of NUREG-0654/FEMA-REP-1, section 50.47(b)(7). Methods or solutions that differ from those described in this supplement may be deemed acceptable if they provide sufficient basis and information for the U.S. Nuclear Regulatory Commission (NRC) staff to verify that the proposed alternative demonstrates compliance with the appropriate NRC regulations. Current licensees may continue to use guidance the NRC found acceptable for complying with 10 CFR 50.47(b)(7) and (10) as long as their current licensing basis remains unchanged.

Licensees may use the information in this supplement for actions that do not require NRC review and approval such as changes to an emergency plan under 10 CFR 50.54 that do not require prior NRC review and approval. Licensees may use the information in this supplement to resolve regulatory or inspection issues.

---

[1] Considering the possibility that an intentional hostile action directed at a nuclear reactor site could become the direct cause of a radiological emergency, use of the term "incident" is more prudent than "accident" in general discussions on radiological planning and preparedness. With the exception of referencing previous sources, this guidance uses the term "incident," irrespective of initiating events for emergency scenarios.

[2] "Licensees" refers to licensees of nuclear power plants under 10 CFR Parts 50 and 52, and the term "applicants" refers to applicants for licenses for nuclear power plants under 10 CFR Parts 50 and 52 and all applicants for early site permits with complete and integrated emergency plans submitted under 10 CFR Part 52.

[3] In this section, "voluntary" and "voluntarily" means that the licensee is seeking the action of its own accord, without the force of a legally binding requirement or an NRC representation of further licensing or enforcement action.

During regulatory discussions on plant specific operational issues, the NRC staff may discuss with licensees various actions consistent with this supplement, as one acceptable means of meeting Appendix E to Part 50, Section IV, paragraph 3, the applicable provisions of 10 CFR 50.47(b)(10), and, when used in conjunction with related portions of NUREG-0654/FEMA-REP-1, 10 CFR 50.47(b)(7). Such discussions would not ordinarily be considered backfitting even if prior versions of this supplement are part of the licensing basis of the facility. However, unless this supplement is part of the licensing basis for a facility, the NRC staff may not represent to the licensee that the licensee's failure to comply with the guidance in this supplement constitutes a violation.

If an existing licensee voluntarily seeks a license amendment or change and (1) the NRC staff's consideration of the request involves a regulatory issue directly relevant to this supplement and (2) the specific subject matter of this supplement is an essential consideration in the NRC staff's determination of the acceptability of the licensee's request, then the staff may request that the licensee either follow the guidance in this supplement or provide an equivalent alternative process that demonstrates compliance with the underlying NRC regulatory requirements. This is not considered backfitting as defined in 10 CFR 50.109(a)(1) or a violation of any of the issue finality provisions in 10 CFR Part 52.

The NRC staff does not intend or approve any imposition or backfitting of the guidance in this supplement related to implementing Appendix E to Part 50, Section IV, paragraph 3 and 10 CFR 50.47(b)(7) and (10). The NRC staff does not expect any existing licensee to use or commit to using this guidance, unless the licensee makes a change to its licensing basis. The NRC staff does not expect or plan to request licensees to voluntarily adopt this guidance to resolve a generic regulatory issue. The NRC staff does not expect or plan to initiate NRC regulatory action which would require the use of this guidance. Examples of such unplanned NRC regulatory actions include issuance of an order requiring the use of this guidance, requests for information under 10 CFR 50.54(f) as to whether a licensee intends to commit to use of this guidance, generic communication, or promulgation of a rule requiring the use of this guidance without further backfit consideration.

Additionally, an existing applicant may be required to adhere to new rules, orders, or guidance if 10 CFR 50.109(a)(3) applies.

The regulation at 44 CFR 350.5(10) states, in part, "Guidelines for the choice of protective actions during an emergency, consistent with Federal guidance, are developed and in place...." Offsite response organizations (OROs) should consider using this guidance to develop appropriate protective action strategies. Nothing in this guide should be interpreted as interfering with State, Tribal, and local ORO authority to determine the appropriate protective action strategies and decision making protocols for the protection of public health and safety during a radiological emergency.

1.3 Protective Action Guides

This guidance should not affect the use of the protective action guidelines developed and issued by the U.S. Environmental Protection Agency (EPA). The EPA protective action guides (EPA, 1992) remain the appropriate Federal guidance on radiological criteria for consideration of protective actions.

1.4 Background

The Federal Emergency Management Agency (FEMA) and the NRC issued Supplement 3 to NUREG-0654/FEMA-REP-1 in 1996 (NRC/FEMA, 1996) as a draft report for interim use and comment. At that time, the guidance reflected the most current accident analysis information (NUREG-1150, "Severe Accident Risks: An Assessment for Five U.S. Nuclear Power Plants," issued December 1990 (NRC, 1990)). The 1996 version of Supplement 3 noted that the guidance was to be used to develop PARs in response to severe accidents. In practice, this was translated into the expectation that protective actions would be recommended and implemented during any General Emergency. Although a General Emergency is a serious event and warrants protective action, it is not necessarily synonymous with a "severe accident" as that term is used in nuclear power plant accident analyses. The current guidance recognizes the disparity between a severe accident with early release and other General Emergency conditions and has provided scenario-specific protective action decision guidance. Additionally, it provides guidance for the immediate evacuation of those closest to the nuclear plant and criteria for the expansion of initial protective actions. This is intended to simplify initial protective action decision making and implementation and allow additional time for evacuation management if the expansion of evacuated areas is necessary.

In late 2004, the NRC initiated a project to analyze the relative efficacy of alternative protective action strategies in reducing consequences to the public from a spectrum of nuclear power plant core melt accidents. The study is documented in NUREG/CR-6953, "Review of NUREG-0654, Supplement 3, 'Criteria for Protective Action Recommendations for Severe Accidents,'" Volumes 1, 2, and 3 (NRC, 2007a; NRC, 2008; and NRC, 2010) (hereinafter referred to as the PAR study). The PAR study provides a technical basis for enhancing protective action guidance.

The NRC Advisory Committee on Reactor Safeguards reviewed the PAR study and documented its review in a July 27, 2007, letter to the NRC (NRC, 2007b), recommending a revision of NUREG-0654/FEMA-REP-1, Supplement 3. In 2008, the Commission directed the NRC staff to revise Supplement 3 in coordination with FEMA after receiving public input.

The following summarizes the results of the PAR study for enhancing protective action strategies:

- Radial evacuation should remain the major element of protective action strategies.

  Evacuations are effective in protecting public health and safety, and the public is seldom injured during evacuation. Local authorities successfully perform many evacuations without the benefit of approved evacuation plans.

- Sheltering in place (SIP) should receive more emphasis in protective action strategies.

  SIP is more protective than radial evacuation under rapidly progressing severe accidents at sites with long evacuation times. Direction for SIP should include instructions for those individuals at work or in transit. Authorities must understand how and when to end SIP (ORNL, 2003). Effective SIP can reduce a shadow evacuation.

- Staged evacuation is the preferred initial protective action in response to a General Emergency and should be considered.

Staged evacuation is more protective than immediate radial evacuation in many scenarios because it limits the exposure of those individuals closest to the plant. In some scenarios, the improved benefit is not large; however, in most every case, a staged evacuation speeds decision making, decreases demand on ORO traffic control and reception center resources, simplifies initial protective actions, and reduces public disruption.

- Precautionary actions, such as evacuating schools and parks during a Site Area Emergency, can be prudent.

  A review of Site Area Emergencies since 1980 shows that none required precautionary actions. However, in the more significant scenarios considered in the emergency preparedness planning basis, such actions are prudent to protect public health and safety.

- Strategies that reduce evacuation time also reduce public health consequences.

  Staged evacuation can reduce evacuation times by allowing the early movement of some people while traffic and access control points are being set up to further direct road use. Staged evacuation is most beneficial if shadow evacuation is minimized.

- Evacuation time estimates (ETEs) are important in planning protective action strategies.

  ETEs can be used as timing criteria to expand staged evacuation and for SIP versus evacuation decision making for large early release scenarios. Recognition of the evacuation "tail" has improved planning. The evacuation tail is about the last 10 percent of evacuees who have longer preparation time (e.g., farms and businesses) (Wolshon, 2010). Although evacuation plans account for all members of the public, a decision based on the evacuation time of 90 percent of the population reduces total public exposure.

- Advance planning for the evacuation of persons with disabilities and access/functional needs who do not reside in special facilities may not be consistently addressed within all nuclear power plant emergency planning zones (EPZs).

  The PAR study survey identified that evacuation planning may under serve residents who need assistance from outside the home to evacuate (NRC, 2008). Reasonable enhancements, including the integration of more modern technologies in the registration process, can be made to improve planning for this population.

- Emergency messaging and public information should be improved to reduce shadow evacuation and to improve compliance with protective action direction.

  In order to reduce shadow evacuation, the public must receive frequent information and specific instructions. Under standard offsite conditions (e.g., no severe weather or hostile activities), SIP for all areas not evacuated may not be necessary or appropriate because it would limit people from reuniting with their families (e.g., retrieving children).

These results contributed to the update of Supplement 3 to NUREG-0654/FEMA-REP-1. Input from State and local government emergency response professionals, stakeholders, and industry was also incorporated.

The guidance in Supplement 3 to NUREG-0654/FEMA-REP-1 is based on recent technical analyses, and it updates the following previous guidance on the development of PAR strategies for nuclear power plant accidents:

- Appendix 1, "Emergency Action Level Guidelines for Nuclear Power Plants," to NUREG-0654/FEMA-REP-1, Revision 1, "Criteria for Preparation and Evaluation of Radiological Emergency Response Plans and Preparedness in Support of Nuclear Power Plants," issued November 1980 (NRC/FEMA, 1980)

  Note that the revised Supplement 3 updates the protective action guidance in Appendix 1 to NUREG-0654/FEMA-REP-1, Revision 1 (e.g., for the General Emergency on pages 1-16 through 1-19) but does not address other guidance in Appendix 1.

- Supplement 3, "Criteria for Protective Action Recommendations for Severe Accidents," to NUREG-0654/FEMA-REP-1, Revision 1, published in July 1996 as a draft report for interim use and comment (NRC, 1996)

- NRC Information Notice 83-28, "Criteria for Protective Action Recommendations for General Emergencies," dated May 4, 1983 (NRC, 1983)

- NRC Regulatory Issue Summary 2003-12, "NRC Regulatory Issue Summary 2003-12: Clarification of NRC Guidance for Modifying Protective Actions," dated June 24, 2003 (NRC, 2003)

- Revision 5 to Volume 1 of NUREG/BR-0150, "RTM-96: NRC Response Technical Manual," issued October 2002 (NRC, 2002)

2. PROTECTIVE ACTION STRATEGY GUIDANCE

2.1 Development of Site-Specific Protective Action Recommendation and Decision Logic

The attachment to this supplement contains a protective action strategy development tool that OROs should consider using to develop site-specific protective action strategy guidance for decision makers and that licensees should use to develop PAR procedures. **The protective action strategy development tool in the attachment should not be used without site-specific modification.**

The emergency plan implementing procedures used by the nuclear power plant emergency response organization (ERO) should include a site-specific PAR development tool. The attachment is intended to guide the development of such a tool for operational shift personnel and is designed to be implemented rapidly without the initial need to confer with ORO personnel. The PAR tool used by the licensee-augmented ERO may differ depending on whether the augmented ERO has more resources than the shift organization. Section IV.D.3 of Appendix E, "Emergency Planning and Preparedness for Production and Utilization Facilities," to 10 CFR Part 50, "Domestic Licensing of Production and Utilization Facilities," requires licensees to have the capability to notify OROs within 15 minutes of the declaration of a General Emergency.

As demonstrated in biennial evaluated exercises, licensees include a PAR with the General Emergency notification. The 15-minute notification requirement remains in effect regardless of differences in the licensee PAR tools used by shift and augmented ERO personnel. The PAR must be made rapidly, in accordance with approved procedures, and the licensee should develop those procedures in partnership with the ORO(s) responsible for protective action decision making.

The background information and notes included with the protective action strategy development tool provide direction for developing site-specific elements and criteria. The tool diagram is simplified when the site-specific elements are developed and deployed in an emergency plan implementing procedure. The tool notes some decision criteria that should be discussed and agreed upon by licensees and responsible OROs. However, in no case does the NRC intend that nuclear power plant licensees delay the recommendation of protective actions in order to confer with OROs at the time of a General Emergency. Licensees are responsible for making timely PARs in accordance with regulations, Federal guidance, and plant conditions and for providing the PARs to OROs to allow them to make timely and well-informed protective action decisions. OROs are responsible for deciding which protective actions to implement.

FEMA and the NRC expect nuclear power plant licensees to develop PAR procedures that include ORO input for various decision points. The approved PAR emergency plan implementing procedure constitutes the licensee's commitment to OROs to provide PARs immediately upon the declaration of a General Emergency in a manner mutually agreed upon. In case a responsible ORO chooses not to participate in the development of a site-specific PAR development tool that is consistent with this guidance, the licensee may use FEMA-approved ORO emergency plans and implementing procedures as a basis to develop the necessary decision points. Efforts to achieve licensee and ORO agreement on protective action strategy decision criteria should be documented in a manner that makes the information available for review by the NRC and FEMA.

It is incumbent upon licensees to make the determination that onsite emergency plan changes meet the requirements in 10 CFR 50.54(q).

2.2   Notification of Protective Action Recommendations at a General Emergency

Licensees are required to be able to provide immediate notification (i.e., within 15 minutes) to OROs upon the declaration of a General Emergency. The General Emergency notification is expected to include a PAR. The PAR must be developed in accordance with the approved onsite emergency plan the PAR procedure and should be coordinated with OROs.

2.3   Termination of Protective Actions

Licensees are responsible for declaring a General Emergency and issuing a PAR. The licensee is also responsible for downgrading or terminating the General Emergency; however, it should not take this action without wide consultation. Downgrading an emergency may take time to ensure that the plant condition will remain safe and to confer with authorities. OROs are responsible for terminating offsite protective actions. The licensee provides input on the plant's status to ORO decision makers.

2.4   Precautionary Protective Actions at Site Area Emergency

OROs at many sites plan precautionary actions upon the declaration of a Site Area Emergency. These actions may include sounding sirens, informing the population that an event has taken place, closing schools, closing parks, and preparing special needs facilities for potential evacuation.

In some cases, a licensee or ORO may have committed to site-specific precautionary actions during the Site Area Emergency, such as the evacuation of beaches or other recreational areas. Licensees should not interpret this guidance as countermanding any such commitments in licensing-basis documents or in State, Tribal, and local offsite emergency plans and implementing procedures.

2.5   Wind Persistence Issues

It may be appropriate for licensees to perform a wind persistence analysis (the updated final safety analysis may be used) to determine appropriate modifications to a protective action strategy. The modifications may be appropriate for areas where the typical site meteorology includes wind direction shifts on a timescale that is shorter than the ETE for downwind 2- to 5-mile sectors. This could result in OROs expanding protective actions while an evacuation is in progress as a result of changes in wind direction. Multiple changes in protective action direction may undermine credibility and increase shadow evacuations and thereby increase evacuation times. In such cases, it may be appropriate to include more than three downwind sectors in an expanded evacuation.

2.6   Expansion of Initial Protective Action Recommendations

The emergency action level scheme used at nuclear power plants is designed to be anticipatory. A General Emergency is expected to be declared, based on plant conditions, before a radiological release could potentially begin. Licensees will perform radiological assessments throughout the emergency and will recommend to OROs the need to take or expand protective actions if dose projections show that protective action criteria could be exceeded. Dose projections that are based on effluent monitor data and verified by field monitoring data would provide the strongest basis for a PAR; however, effluent monitor data alone can be sufficient if other data (e.g., plant conditions, area or process monitors) verify the occurrence of a radiological release. Although verification of dose projection data is desirable, the licensee should not delay PARs unduly while waiting for field monitoring data or sample analysis.

A more difficult case for dose assessment is a scenario with a large radiological source term in containment and a leak rate at or near the design basis. This is clearly a General Emergency and an initial PAR is expected. As subsequent PARs are implemented, the issue of expansion of protective actions beyond the 5-mile downwind sectors can arise. When expansion of a PAR is considered under this scenario, the condition of containment must be assessed.

Additionally, changes in wind direction may indicate that if a release begins, it would affect different downwind sectors. If the licensee believes that containment may fail, it should pursue the expansion of PARs. Finally, if a radiological assessment shows that an ongoing release or containment source term is not sufficient to cause exposures in excess of EPA protective action guidelines, licensees should not expand PARs based only on changes in wind direction.

2.7   Strategy for Rapidly Progressing Scenarios

The emergency preparedness planning basis includes rapidly progressing scenarios that have a significant radioactive release in about 1 hour. Historically, emergency preparedness regulations and guidance have been based on a spectrum of accidents. NUREG-0396, "Planning Basis for the Development of State and Local Government Radiological Emergency Response Plans in Support of Light-Water Nuclear Power Plants," issued November 1978, embodies this concept in the specification of the EPZ (NRC, 1978). Furthermore, NUREG-0654/FEMA-REP-1, Revision 1, notes that planning should not address a single accident sequence as each accident could have different consequences (NRC/FEMA, 1980).

To provide a technical basis for the development of PARs for a rapidly progressing scenario, the NRC staff performed a series of calculations using a spectrum of source terms (NRC, 2010). The objective was to identify the relative efficacy of protective action options at sites with differing population densities. The analysis included SIP, evacuation at different distances, varied shelter durations, and evacuation speeds. The analysis evaluated the efficacy of protective actions for the 0- to 2-mile, 2- to 5-mile, and 5- to 10-mile zones.

Factors that most influenced the efficacy of protective action strategies included the travel speed of the evacuating population and shelter duration. Travel speed is related to population density and is influenced by the roadway network and evacuation planning. The analysis derived the travel speeds from current estimates for evacuating 90 percent of the general public under normal weekday conditions (NRC, 2010). The analysis tested multiple weather trials and assessed mean consequences. The calculations determined relative efficacy rather than absolute consequences.

For sites at which the 90-percent ETE for the general public of the full EPZ is less than approximately 3 hours, results showed that, for the rapidly progressing scenario, evacuation is the most appropriate protective action. For sites where this is not the case, the protective actions listed in the table below are most beneficial, unless impediments exist to their implementation. Where evacuation cannot be accomplished in the time specified, a recommendation for SIP until the plume has passed is more beneficial. The evacuation tail generally represents the last 10 percent of the population and describes the population that takes a disproportionately longer time to evacuate than the remaining public. Planning is in place to evacuate 100 percent of the public; however, PARs and decisions should be based on the 90-percent ETE values.

ZONE	PROTECTIVE ACTION
0 to 2 mile	If the 90-percent ETE for this area is 2 hours or less, immediately evacuate.
2 to 5 mile	If the 90-percent ETE for this area is 3 hours or less, immediately evacuate.
5 to 10 mile	SIP, then evacuate when it is safe to do so.

Licensees may perform a site-specific analysis to determine whether other criteria are more appropriate.

Extreme weather conditions, such as inversion, significant precipitation, or no wind, can change the efficacy of SIP and make evacuation the preferred protective action.

3. PUBLIC INFORMATION MATERIALS

The public information program required by 10 CFR 50.47(b)(7) is intended to provide the permanent and transient population within the EPZ the opportunity to become aware of preparedness information (NRC, 1980). In a 2008 telephone survey of residents in EPZs published in NUREG/CR-6953, Volume 2 (NRC, 2008), most respondents stated that they are familiar with these emergency information materials, and many of them keep this information readily accessible. NUREG-0654/FEMA-REP-1, Revision 1, contains guidance on public information materials and notification of the public. An informed EPZ population contributes to reducing evacuation times by doing the following:

- reducing preparation time

- reducing the time to accept and implement the protective action direction (sometimes called the "milling" time)

- reducing shadow evacuation

- reducing travel times through a knowledge of evacuation routes

An informed population also enhances the implementation of SIP strategies.

The guidance in this section is meant to enhance public information for EPZ populations and improve messaging during emergencies to enhance public compliance with protective action direction.

3.1 Public Information

Public information materials should describe protective action strategies, including staged evacuation. The materials must also stress the need for the public to monitor information channels for updated direction.

If OROs use this strategy, public information materials should explain that the purpose of staged evacuation is to allow those closest to the plant to evacuate unimpeded. Materials should clearly state that those individuals who are not within the declared evacuation area should not evacuate. The materials should define the term "shadow evacuation" and note that this type of evacuation has the potential to impede the outbound traffic flow, slowing the evacuation from the affected area.

ORO expectations for those under an advisory to monitor and prepare should be stated (e.g., reunite with family members, prepare for evacuation, monitor information channels, and keep off the road).

Public information materials tend to be directed to individuals who are at home when an emergency occurs. Materials should clarify expectations for those who are not at home when a protective action is ordered.

Instructions should include how to SIP and provide details such as closing doors and windows, turning off air conditioning or heating (as appropriate), and monitoring communications channels for further instructions. The materials should also include information for people in vehicles when an SIP order is issued (e.g., leave the EPZ or enter a nearby building) and for people who are away from home, such as working, shopping, or dining (e.g., remain in the building where they are currently located and monitor for additional information).

If the ORO program includes potassium iodide (KI), instructions should include what to do if KI is not available (e.g., residents cannot find KI, or they do not have it with them). Residents must understand that KI is an additional precaution, and that they still should be safe if KI has been recommended but is inaccessible.

Through cell phone use among school age children, parents will likely become aware of an impending school evacuation before buses are mobilized. This early awareness may result in some parents picking up their children. Materials should discuss the benefit of allowing schools to implement school evacuation plans without interference.

A significant lesson learned following the Hurricane Katrina disaster is that planning and preparedness for special needs populations was lacking and needed to be made more robust and enhanced. The National Response Framework defines "special needs populations" as "populations whose members may have additional needs before, during, and after an incident in functional areas, including but not limited to: maintaining independence, communication, transportation, supervision and medical care. Individuals in need of additional response assistance may include; those who have disabilities; who live in institutionalized settings; who are elderly; who are children; who are from diverse cultures; who have limited English proficiency or are non-English speaking; or who are transportation disadvantaged."

Following the guidance provided by FEMA's Office of Disability Integration and Coordination and other responsible departments, agencies, and organizations, radiological emergency preparedness (REP) OROs (working in a partnering and closely coordinated manner with FEMA) should emphasize major functional needs as vital to protecting life and safety. This can be accomplished by providing populations with disabilities and access/functional needs with functional needs support services (FNSSs), defined as "services that are provided to individuals during an emergency in general population shelters or other integrated community facilities to enable them to maintain their independence in such settings." Historically, REP program public information materials have focused on urging residents with disabilities and access/functional needs to register for assistance by filling out and mailing in registration cards. Results of a national telephone survey of EPZ residents show that 6 percent of the EPZ population may be residents with disabilities and access/functional needs who do not reside in institutionalized settings (NRC, 2008). Of that 6 percent, less than one-third has registered with local authorities, although most residents stated they were willing to inform authorities of their need. The spectrum of preparedness efforts should be expanded and should not be limited to any single method, such as filling out registration cards. The use of community-driven integrated process teams to brainstorm, prioritize, design, and implement FNSSs for their citizens with disabilities and access/functional needs has proven to be quite effective and should be considered.

3.2 Emergency Messaging

The following guidance addresses communication elements that can enhance public compliance with ORO direction to monitor and prepare, SIP, or evacuate.

The alert and notification system warns the public of an emergency and of the need to take protective actions. Emergency Alert System (EAS) messages and radio and television broadcasts communicate information to the public. Information can also be transmitted using local real-time internet postings, text messaging, and other methods consistent with State, Tribal, and local emergency plans.

After the initial alert and notification, the public will maintain an awareness of the event through media broadcasts and subsequent EAS messages. OROs are encouraged to use supplemental information bulletins to inform the public of status and direction periodically.

Information should cover topics such as the following:

- the length of time that members of the public will be expected to SIP and that they will be told to evacuate should it be necessary

- the appropriate actions for SIP in a residence and when in transit

- the projected length of time necessary for the public to monitor and prepare

- a request that people stay off the road to reduce shadow evacuation

- what people should do if they are ordered to evacuate

- who is to evacuate

- where evacuees are to go

- when evacuees need to leave

- transportation alternatives

- instructions for persons with disabilities and access/functional needs who are not in special facilities

Messaging directed to the transit-dependent population should emphasize the need to request a ride from a neighbor, relative, or friend. For those who cannot obtain a ride, information should be provided about bus routes, how residents are expected to get to the bus route, and what to do if they cannot get to a bus route. Additionally, information should explain why they will be safe outdoors while waiting to be picked up, what they should bring, and how long they may expect to wait for a bus. This population group may include thousands of individuals.

Messaging to persons with disabilities and access/functional needs who do not reside in special facilities should request that these residents obtain a ride from a relative, friend, or neighbor, if at all possible. Instructions should clearly explain to residents who have previously registered with authorities how long they should expect to wait for prearranged assistance, what to do while they are waiting, and what to do if assistance does not arrive during the specified timeframe. Mobilizing and completing an evacuation in some EPZs can take many hours. Instructions are also needed for residents who have not preregistered but who require transportation assistance.

Special facilities have specific evacuation plans and typically would receive early warning through direct notification from OROs. This preplanned activity helps ensure that the special facilities are notified promptly to allow response activities to begin.

4. GLOSSARY

Emergency Response Planning Area. A local area within the EPZ for which emergency response information is provided (NRC, 2005). These areas are typically defined by geographic or political boundaries to support emergency response planning and may not conform to an exact 10-mile (16-kilometer) radius from the nuclear power plant.

Evacuation Tail. A small portion of the population that takes a disproportionately longer amount of time to evacuate than the remaining public and is the last to leave the evacuation area. The tail generally consists of approximately the last 10 percent of the population.

Evacuation Time Estimate (ETE). The estimated time needed to evacuate the public from affected areas of the plume exposure pathway EPZ.

Monitor and Prepare. A type of precautionary action intended to advise the public within the EPZ that a serious emergency at the nuclear power plant exits and that it should monitor the situation and prepare for the possibility of evacuation, SIP, or other protective actions. Further, if an evacuation is underway, officials should ask individuals who are not involved in the evacuation to remain off the roadways to allow those who are instructed to evacuate to do so.

Shelter in Place (SIP). A type of protective action in which instructions are given to members of the public to remain indoors, turn off heating or air conditioning (as appropriate for the region and season), close windows, monitor communication channels, and prepare to evacuate. Those individuals who are not at home (e.g., shopping, dining, working) are instructed to stay in their current location. The instructions should specify that SIP is safer than evacuation at this time, or that, alternatively, the SIP is being implemented to ensure that the public remain off roadways to allow other areas that are under an evacuation order to evacuate unimpeded. The intent is that members of the public should remain where they are or should seek shelter close by, but they should not return home to shelter.

Emergency Planning Zone (EPZ). A geographic area surrounding a commercial nuclear power plant for which emergency planning is necessary to ensure that OROs can take prompt and effective actions to protect public health and safety in the event of a radiological accident. The plume pathway EPZ is approximately 10 miles in radius, whereas the ingestion pathway EPZ has a radius of approximately 50 miles.

Persons with Disabilities and Access/Functional Needs. Individuals within a community that may have additional needs before, during, and after an incident in one or more of the following functional areas: (1) maintaining independence, (2) communication, (3) transportation, (4) supervision, and (5) medical care. Individuals who are in need of additional response assistance may include those who have sensory, motor skill, or mental/emotional disabilities, who live in institutionalized settings, who are elderly, who are children, who are from diverse cultures, who have limited or no English-speaking proficiency, or who are transportation-disadvantaged.

Shadow Evacuation. The spontaneous evacuation of people from areas outside of the official evacuation zone.

5. REFERENCES

10 CFR Part 50. *Code of Federal Regulations*, Title 10, *Energy,* Part 50, "Domestic Licensing of Production and Utilization Facilities."

44 CFR Part 350. *Code of Federal Regulations*, Title 44, *Emergency Management and Assistance,* Part 350, "Review and Approval of State and Local Radiological Emergency Plans and Preparedness."

Oak Ridge National Laboratory (ORNL). 2003. ORNL/TM-2003/230, "Questions and Answers regarding Actions To Take When Ending Shelter in Place." U.S. Department of Homeland Security Chemical Stockpile Emergency Preparedness Program Protective Action Working Integrated Process Team. Oak Ridge, TN. September 2003.

U.S. Environmental Protection Agency (EPA). 1992. EPA-400-R-92-001, "Manual of Protective Action Guides and Protective Actions for Nuclear Incidents." Washington, DC. May 1992.

U.S. Nuclear Regulatory Commission (NRC). 2010. NUREG/CR-6953, "Review of NUREG-0654, Supplement 3, 'Criteria for Protective Action Recommendations for Severe Accidents—Technical Basis for Protective Action Logic Diagram,'" Volume 3. Washington, DC. August 2010.

U.S. Nuclear Regulatory Commission (NRC). 2008. NUREG/CR-6953/SAND2008-4195P, "Review of NUREG-0654, Supplement 3, 'Criteria for Protective Action Recommendations for Severe Accidents—Focus Groups and Telephone Survey,'" Volume 2. Washington, DC. October 2008.

U.S. Nuclear Regulatory Commission (NRC). 2007a. NUREG/CR-6953/SAND2007-5448P, "Review of NUREG-0654, Supplement 3, 'Criteria for Protective Action Recommendations for Severe Accidents,'" Volume 1. Washington, DC. December 2007.

U.S. Nuclear Regulatory Commission (NRC). 2007b. Advisory Committee on Reactor Safeguards report entitled, "Draft NUREG/CR, Review of NUREG-0654, Supplement 3, 'Criteria for Protective Action Recommendations for Severe Accidents.'" Washington, DC. July 27, 2007. (Agencywide Documents Access and Management System Accession No. ML071980087)

U.S. Nuclear Regulatory Commission (NRC). 2005. NUREG/CR-6863, SAND2004-5900. "Development of Evacuation Time Estimate Studies for Nuclear Power Plants." Washington D.C. January 2005.

U.S. Nuclear Regulatory Commission (NRC). 2003. NRC Regulatory Issue Summary 2003-12, "Clarification of NRC Guidance for Modifying Protective Actions." Washington, DC. June 24, 2003.

U.S. Nuclear Regulatory Commission (NRC). 2002. NUREG/BR-0150, "RTM-96: Response Technical Manual," Volume 1, Revision 5. Washington, DC. October 2002.

U.S. Nuclear Regulatory Commission (NRC)/Federal Emergency Management Agency (FEMA). 1996. NUREG-0654/FEMA-REP-1, "Criteria for Protective Action Recommendations for Severe Accidents,'" Supplement 3. Washington, DC. July 1996.

U.S. Nuclear Regulatory Commission (NRC). 1990. NUREG-1150, "Severe Accident Risks: An Assessment for Five U.S. Nuclear Power Plants," Washington, DC. December 1990.

U.S. Nuclear Regulatory Commission (NRC). 1983. NRC Information Notice 83-28, "Criteria for Protective Action Recommendations for General Emergencies." Washington, DC. May 4, 1983.

U.S. Nuclear Regulatory Commission (NRC)/Federal Emergency Management Agency (FEMA). 1980. NUREG-0654/FEMA-REP-1, "Criteria for Preparation and Evaluation of Radiological Emergency Response Plans and Preparedness in Support of Nuclear Power Plants," Revision 1. Washington, DC. November 1980.

U.S. Nuclear Regulatory Commission (U.S.) (NRC). 1978. NUREG-0396, "Planning Basis for the Development of State and Local Government Radiological Emergency Response Plans in Support of Light-Water Nuclear Power Plants." Washington, DC. November 1978.

Wolshon, B., J. Jones, and F. Walton. 2010. "The Evacuation Tail and Its Effect on Evacuation Decision Making." *Journal of Emergency Management*. Volume 8, No. 1, January/February 2010.

ATTACHMENT: PROTECTIVE ACTION STRATEGY DEVELOPMENT TOOL

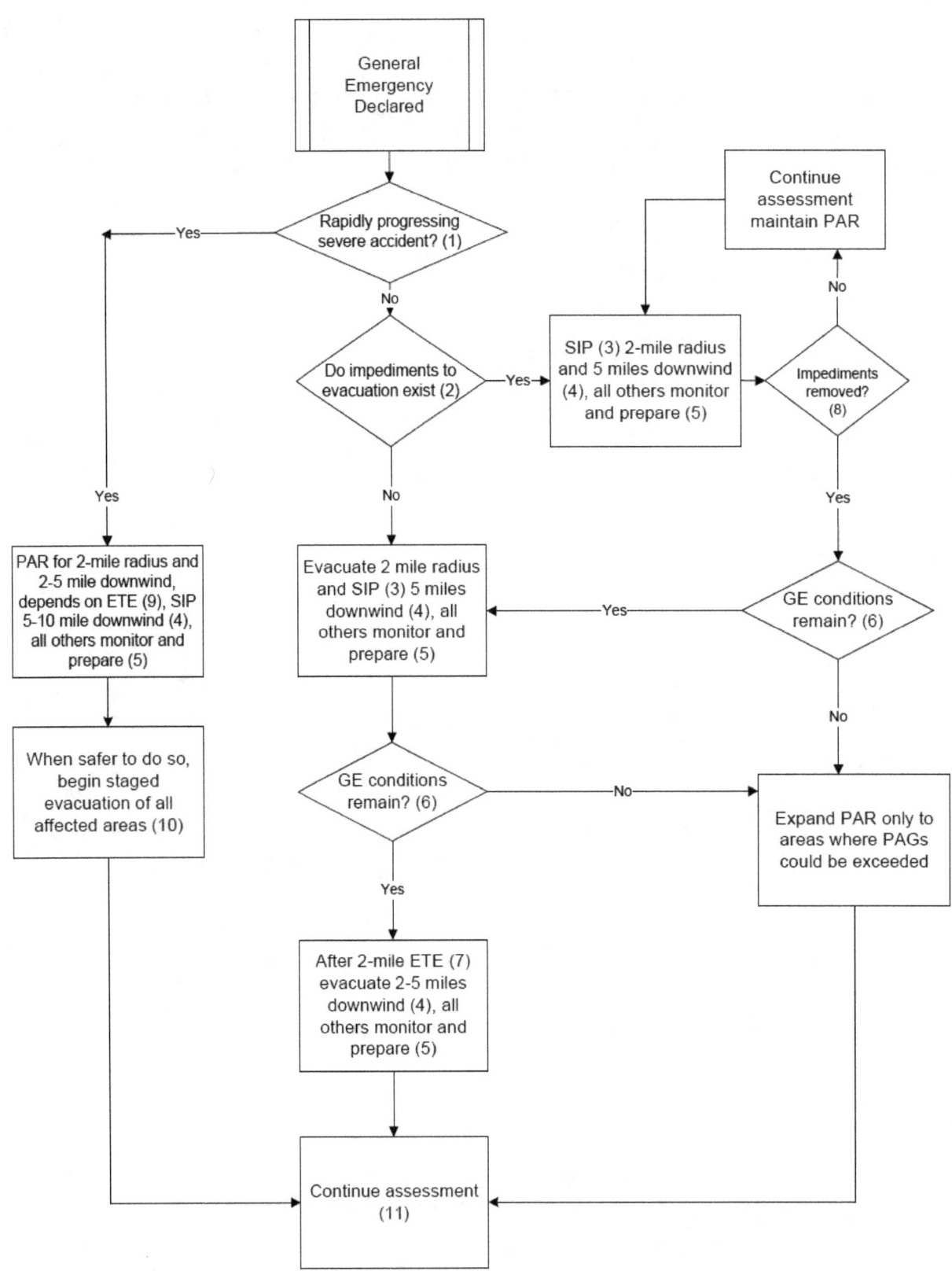

Protective action strategy development tool

PROTECTIVE ACTION STRATEGY DEVELOPMENT TOOL NOTES

It is not intended that licensees or offsite response organizations (OROs) have protective action implementing procedures that are exactly the same as provided here. Licensees for each nuclear power plant should develop site-specific strategies and decision tools/procedures for the site using the guidance provided below in collaboration with OROs responsible for protective action decision making.

The information in these notes that should be considered in developing the strategy is labeled as "Note." Background information is labeled as "Background Note" and is meant to be helpful in development efforts.

Note 1: Rapidly Progressing Severe Incident

A rapidly progressing severe incident is a General Emergency (GE) with rapid loss of containment integrity (emergency action levels indicate containment barrier loss) and loss of ability to cool the core. This path is used for scenarios in which containment integrity can be determined as bypassed or immediately lost during a GE with core damage. If this scenario cannot be immediately confirmed, assume it is not taking place and answer "no" to this decision block.

Note 2: Impediments to Evacuation

Impediments to evacuation include the following:

- Evacuation support (e.g., traffic control) is not yet in place. In this situation, the GE is the initial notification, or if a previous notification was made, the GE notification occurs before preparations to support an evacuation are complete. Many sites have a low population density within 2 miles, and lack of traffic control may not be considered an impediment. The licensee and OROs should discuss this element and reach an agreement. The licensee and OROs should agree, in advance, on an expected time for evacuation support to be put in place after notification of an emergency classification. The site-specific protective action recommendation (PAR) procedure for those sites at which a delay of an initial staged evacuation is necessary, pending support setup, should include this time. The licensee would base procedures on the agreement and would not confer with OROs before making the initial PAR notification.

- In a hostile-action-based GE (armed attack), OROs may determine that an initial recommendation to shelter in place (SIP) rather than evacuation is the preferred path. The licensee would discuss this element with OROs and reach an agreement during the development process. The licensee would base procedures on the agreement and would not confer with OROs before making the initial PAR notification.

- In the event of adverse weather, licensees are not responsible for soliciting information or for making a determination that weather or other impediments (e.g., an earthquake or wildfire) for safe public evacuation exist at the time of the emergency. However, the licensee will consider an impediment to exist if OROs have previously notified it of such an impediment (e.g., roadways are closed because of deep snow). During the planning process, OROs may determine that the licensee does not need to consider adverse weather in its plant PAR procedures.

Note 3: Shelter in Place

SIP means that instructions are given to members of the public to remain indoors, turn off heating or air conditioning (as appropriate for the region and season), close windows, monitor communications channels, and prepare to evacuate. The instructions should specify that SIP is safer than evacuation at this time, or that, alternatively, SIP is being implemented in order to keep roadways clear to allow others to evacuate rapidly. The intent of SIP is for members of the public to remain where they currently are or to seek shelter close by, but they should not return home to shelter when more immediate options for sheltering are available.

Note 4: Downwind Sectors

Downwind sectors include a downwind 22.5-degree compass sector(s) and adjacent sectors. Generally, the downwind sectors involve three or four sectors and include all the emergency response planning areas impacted in that area.

Background Note: Wind Persistence

Site-specific wind persistence information may indicate the need to include additional sectors with the initial recommendation. However, the licensee should discuss this element with responsible OROs to determine whether expanded initial protective actions are appropriate or desirable. The size of emergency response planning areas may determine whether there is a site-specific need for this contingency.

Note 5: Monitor and Prepare

The instruction to monitor and prepare is intended to engage the population within the plume exposure pathway emergency planning zone, inform them of the emergency, and advise them that they should monitor the situation and prepare for the possibility of evacuation, SIP, or other protective actions. If an evacuation is underway, officials should ask members of the public who are not directed to evacuate to remain off the roadways to allow the evacuation to proceed.

Background Note: Emergency Messaging

Effective emergency messaging requires clear and frequent communications with the public. If the public is not engaged (i.e., given instructions of some kind), a larger shadow evacuation could result. A large shadow evacuation could impede those closest to the plant and increase public exposure. Frequent communication may also reduce public inquiries to OROs for status and instructions.

Note 6: Consideration of Plant Conditions before the Evacuation of Downwind Sectors

If the plant has mitigated the conditions that caused the GE declaration (i.e., core cooling is restored), expanding the PAR to evacuate downwind sectors upon completion of the initial staged evacuation may not be necessary. However, if GE emergency action levels are still met, expansion of the PAR to the downwind sectors may be appropriate. If the plant restores core cooling, it must still perform a radiological assessment to identify the extent of contamination, if any. If surveys or dose projections reveal areas under no protective action direction where protective action guidelines (PAGs) could be exceeded, the members of the public in those areas should be evacuated or sheltered, as appropriate.

Note 7: Timing for Evacuation of Downwind Sectors

Implementation of this element should occur at the time of the site-specific 2-mile evacuation time estimate (ETE) for 90-percent evacuation (e.g., T hours (use site-specific time) after OROs were notified of the initial PAR to evacuate downwind sectors).

Background Note: T Values

The licensee will identify the value of T using the site-specific ETE and should consider T_D for a daytime ETE and T_N for a nighttime ETE. These values should be representative for the site and should not include special events (e.g., temporary offsite activities that draw into the emergency planning zone transient, nonresident individuals who may be present during an emergency). However, OROs should consider the effects of special events. If the shift staff is responsible for making this PAR, it should do so without conferring with OROs and in accordance with procedures, based on the ETE value alone. The verification of the evacuation progress is not expected. However, if the augmenting emergency response organization (ERO) has been activated, sufficient resources may be available for the licensee to confer with OROs more fully before expanding the PAR to downwind sectors.

Note 8: Removal of Evacuation Impediments

Removal of evacuation impediments involves the following:

- Evacuation Support. If the OROs identified this contingency as necessary during the planning effort, the licensee should notify OROs with an evacuation PAR when the agreed upon time (e.g., 1 hour from the GE notification) has elapsed. The licensee shift staff is not expected to confer with OROs before changing the PAR, but if the augmenting ERO is activated they may do so.

- Hostile Action (Armed Attack). OROs may identify this contingency as necessary during the planning effort. It may be appropriate to set up a timeframe for the licensee to notify OROs with an evacuation PAR. The licensee shift staff is not expected to confer with OROs before changing the PAR, but if the augmenting ERO is activated they may do so.

- Adverse Weather. If weather or some other roadway disruption caused the impediment, OROs will determine when it is appropriate to change the protective action. Licensees have no responsibility for PAR modification unless a PAR change is necessary because of plant conditions or radiological assessment. OROs determine when it is safe for the public to evacuate.

Note 9: SIP versus Evacuation PAR for Rapidly Progressing Scenarios

The licensee should issue an evacuation PAR in scenarios for which the time to evacuate 90 percent of the population within a 2-mile radius is 2 hours or less. If the ETE is longer, the licensee should recommend SIP. The licensee should consider T_D for a daytime ETE and T_N for a nighttime ETE.

The licensee should issue an evacuation PAR in scenarios for which the 2- to 5-mile downwind sector evacuation time for 90-percent completion is 3 hours or less. If the ETE is longer, the licensee should recommend SIP.

For all cases, the licensee should recommend SIP for the 5- to 10-mile downwind sectors.

To the extent practical and recognizing the urgency of the incident, impediments may be considered. The existence of impediments could change the most effective PAR from evacuation to SIP.

Background Note: Rapidly Progressing Scenario

The ETE values should be representative for the site and should not include special events.

The rapidly progressing incident is more severe than other GEs, and different protective actions are appropriate for all sites.

Extreme weather conditions, such as inversion, significant precipitation, or no wind, can change the efficacy of SIP and make evacuation the preferred protective action.

Licensees may perform an analysis to determine site-specific ETE criteria instead of using this generic guidance.

Note 10: Evacuation Timing for Rapidly Progressing Scenarios

Evacuation after the SIP period is critical for reducing public exposure. Licensees should discuss the evacuation of the sheltered population with OROs.

Background Note: Evacuation Timing for Rapidly Progressing Scenarios

The evacuation should proceed from the areas that are most at risk. The evacuation may involve a 2-mile radius unless field monitoring data show otherwise (e.g., at a site with an elevated release point where contamination may begin beyond 2 miles). Lateral evacuation (e.g., travel perpendicular to the direction of the plume) may be considered where the roadway network is conducive, as it may reduce public exposure. However, preplanning for lateral evacuation is not expected. In any case, the determination of evacuation routes and timing should be based on release information, field monitoring data, and ORO resources.

Note 11: Continue Assessments

Radiological and meteorological assessments should be continued and evacuation considered for any areas where dose projections or field measurements indicate that PAGs may be exceeded.

Background Note: Continue Assessments

Communications with the public should be maintained while protective actions are in effect.

NRC FORM 335 (12-2010) NRCMD 3.7 BIBLIOGRAPHIC DATA SHEET (See instructions on the reverse)	U.S. NUCLEAR REGULATORY COMMISSION	1. REPORT NUMBER (Assigned by NRC, Add Vol., Supp., Rev., and Addendum Numbers, if any.) NUREG-0654 / FEMA-REP-1, Rev. 1, Supplement 3

www.ingramcontent.com/pod-product-compliance
Lightning Source LLC
Chambersburg PA
CBHW081415170526
45166CB00010B/3347